The Fight For Conservation

Gifford Pinchot

THE FIGHT
FOR CONSERVATION·

By
GIFFORD PINCHOT

NEW YORK
DOUBLEDAY, PAGE & COMPANY
1910

To
L. H.

305601

CONTENTS

INTRODUCTION

The following discussion of the conservation problem is not a systematic treatise upon the subject. Some of the matter has been published previously in magazines, and some is condensed and rearranged from addresses made before conservation conventions and other organizations within the past two years.

While not arranged chronologically, yet the articles here grouped may serve to show the rapid, virile evolution of the campaign for conservation of the nation's resources.

I am indebted to the courtesy of the editors of *The World's Work, The Outlook*, and of *American Industries* for the use of matter first contributed to these magazines.

THE FIGHT
FOR CONSERVATION

CHAPTER I

PROSPERITY

THE most prosperous nation of to-day is the United States. Our unexampled wealth and well-being are directly due to the superb natural resources of our country, and to the use which has been made of them by our citizens, both in the present and in the past. We are prosperous because our forefathers bequeathed to us a land of marvellous resources still unexhausted. Shall we conserve those resources, and in our turn transmit them, still unexhausted, to our descendants?

Unless we do, those who come after us will

have to pay the price of misery, degradation, and failure for the progress and prosperity of our day. When the natural resources of any nation become exhausted, disaster and decay in every department of national life follow as a matter of course. Therefore the conservation of natural resources is the basis, and the only permanent basis, of national success. There are other conditions, but this one lies at the foundation.

Perhaps the most striking characteristic of the American people is their superb practical optimism; that marvellous hopefulness which keeps the individual efficiently at work. This hopefulness of the American is, however, as short-sighted as it is intense. As a rule, it does not look ahead beyond the next decade or score of years, and fails wholly to reckon with the real future of the Nation. I do not think I have often heard a forecast of the growth of our population that extended beyond a total of two hundred millions, and that only as a distant

[4]

and shadowy goal. The point of view which this fact illustrates is neither true nor far-sighted. We shall reach a population of two hundred millions in the very near future, as time is counted in the lives of nations, and there is nothing more certain than that this country of ours will some day support double or triple or five times that number of prosperous people if only we can bring ourselves so to handle our natural resources in the present as not to lay an embargo on the prosperous growth of the future.

We, the American people, have come into the possession of nearly four million square miles of the richest portion of the earth. It is ours to use and conserve for ourselves and our descendants, or to destroy. The fundamental question which confronts us is, What shall we do with it?

That question cannot be answered without first considering the condition of our natural resources and what is being done with them to-day. As a people, we have

[5]

been in the habit of declaring certain of
our resources to be inexhaustible. To no
other resource more frequently than coal
has this stupidly false adjective been applied.
Yet our coal supplies are so far from being
inexhaustible that if the increasing rate of
consumption shown by the figures of the
last seventy-five years continues to prevail,
our supplies of anthracite coal will last but
fifty years and of bituminous coal less than
two hundred years. From the point of view
of national life, this means the exhaustion
of one of the most important factors in our
civilization within the immediate future.
Not a few coal fields have already been
exhausted, as in portions of Iowa and Mis-
souri. Yet, in the face of these known
facts, we continue to treat our coal as though
there could never be an end of it. The
established coal-mining practice at the present date does not take out more than one-
half the coal, leaving the less easily mined
or lower grade material to be made perma-

[6]

nently inaccessible by the caving in of the abandoned workings. The loss to the Nation from this form of waste is prodigious and inexcusable.

The waste in use is not less appalling. But five per cent. of the potential power residing in the coal actually mined is saved and used. For example, only about five per cent. of the power of the one hundred and fifty million tons annually burned on the railways of the United States is actually used in traction; ninety-five per cent. is expended unproductively or is lost. In the best incandescent electric lighting plants but one-fifth of one per cent. of the potential value of the coal is converted into light.

Many oil and gas fields, as in Pennsylvania, West Virginia, and the Mississippi Valley, have already failed, yet vast amounts of gas continue to be poured into the air and great quantities of oil into the streams. Cases are known in which great volumes of oil were systematically burned in order to get rid of it.

The prodigal squandering of our mineral fuels proceeds unchecked in the face of the fact that such resources as these, once used or wasted, can never be replaced. If waste like this were not chiefly thoughtless, it might well be characterized as the deliberate destruction of the Nation's future.

Many fields of iron ore have already been exhausted, and in still more, as in the coal mines, only the higher grades have been taken from the mines, leaving the least valuable beds to be exploited at increased cost or not at all. Similar waste in the case of other minerals is less serious only because they are less indispensable to our civilization than coal and iron. Mention should be made of the annual loss of millions of dollars worth of by-products from coke, blast, and other furnaces now thrown into the air, often not merely without benefit but to the serious injury of the community.

[8]

In other countries these by-products are saved and used.

We are in the habit of speaking of the solid earth and the eternal hills as though they, at least, were free from the vicissitudes of time and certain to furnish perpetual support for prosperous human life. This conclusion is as false as the term "inexhaustible" applied to other natural resources. The waste of soil is among the most dangerous of all wastes now in progress in the United States. In 1896, Professor Shaler, than whom no one has spoken with greater authority on this subject, estimated that in the upland regions of the states south of Pennsylvania three thousand square miles of soil had been destroyed as the result of forest denudation, and that destruction was then proceeding at the rate of one hundred square miles of fertile soil per year. No seeing man can travel through the United States without being struck with the enormous and unnecessary loss of fertility by

[9]

easily preventable soil wash. The soil so lost, as in the case of many other wastes, becomes itself a source of damage and expense, and must be removed from the channels of our navigable streams at an enormous annual cost. The Mississippi River alone is estimated to transport yearly four hundred million tons of sediment, or about twice the amount of material to be excavated from the Panama Canal. This material is the most fertile portion of our richest fields, transformed from a blessing to a curse by unrestricted erosion.

The destruction of forage plants by overgrazing has resulted, in the opinion of men most capable of judging, in reducing the grazing value of the public lands by one-half. This enormous loss of forage, serious though it be in itself, is not the only result of wrong methods of pasturage. The destruction of forage plants is accompanied by loss of surface soil through erosion; by forest destruction; by corresponding deterioration in the water supply; and by a serious decrease in the quality and weight of animals grown on overgrazed lands. These sources of loss from failure to conserve the range are felt to-day. They are accompanied by the certainty of a future loss not less important, for range lands once badly over-grazed can be restored to their former value but slowly or not at all. The obvious and certain remedy is for the Government to hold and control the public range until it can pass into the hands of settlers who will make their homes upon it. As methods of agriculture improve and new dry-land crops are introduced, vast areas once considered unavailable for cultivation are being made into prosperous homes; and this movement has only begun.

The single object of the public land system of the United States, as President Roosevelt repeatedly declared, is the making and maintenance of prosperous homes. That object cannot be achieved unless such of the public lands as are suitable for settlement are conserved for the actual home-maker. Such lands should pass from the possession of the Government directly and only into the hands of the settler who lives on the land. Of all forms of conservation there is none more important than that of holding the public lands for the actual home-maker.

It is a notorious fact that the public land laws have been deflected from their beneficent original purpose of home-making by lax administration, short-sighted departmental decisions, and the growth of an unhealthy public sentiment in portions of the West. Great areas of the public domain have passed into the hands, not of the home-maker, but of large individual or corporate owners whose object is always the making of profit and seldom the making of homes. It is sometimes urged that enlightened self-interest will lead the men who have acquired large holdings of public lands to put

them to their most productive use, and it is
said with truth that this best use is the tillage
of small areas by small owners. Unfortunately, the facts and this theory disagree.
Even the most cursory examination of
large holdings throughout the West will
refute the contention that the intelligent
self-interest of large owners results promptly
and directly in the making of homes. Few
passions of the human mind are stronger
than land hunger, and the large holder
clings to his land until circumstances make
it actually impossible for him to hôld it any
longer. Large holdings result in sheep or
cattle ranges, in huge ranches, in great
areas held for speculative rise in price, and
not in homes. Unless the American home-
stead system of small free-holders is to be so
replaced by a foreign system of tenantry,
there are few things of more importance
to the West than to see to it that the public
lands pass directly into the hands of the
actual settler instead of into the hands of

the man who, if he can, will force the settler to pay him the unearned profit of the land speculator, or will hold him in economic and political dependence as a tenant. If we are to have homes on the public lands, they must be conserved for the men who make homes.

The lowest estimate reached by the Forest Service of the timber now standing in the United States is 1,400 billion feet, board measure; the highest, 2,500 billion. The present annual consumption is approximately 100 billion feet, while the annual growth is but a third of the consumption, or from 30 to 40 billion feet. If we accept the larger estimate of the standing timber, 2,500 billion feet, and the larger estimate of the annual growth, 40 billion feet, and apply the present rate of consumption, the result shows a probable duration of our supplies of timber of little more than a single generation.

Estimates of this kind are almost inevitably misleading. For example, it is certain that the rate of consumption of timber will increase enormously in the future, as it has in the past, so long as supplies remain to draw upon. Exact knowledge of many other factors is needed before closely accurate results can be obtained. The figures cited are, however, sufficiently reliable to make it certain that the United States has already crossed the verge of a timber famine so severe that its blighting effects will be felt in every household in the land. The rise in the price of lumber which marked the opening of the present century is the beginning of a vastly greater and more rapid rise which is to come. We must necessarily begin to suffer from the scarcity of timber long before our supplies are completely exhausted.

It is well to remember that there is no foreign source from which we can draw cheap and abundant supplies of timber to meet a demand per capita so large as to be without parallel in the world, and that the suffering which will result from the progressive failure of our timber has been but faintly foreshadowed by temporary scarcities of coal.

What will happen when the forests fail? In the first place, the business of lumbering will disappear. It is now the fourth greatest industry in the United States. All forms of building industries will suffer with it, and the occupants of houses, offices, and stores must pay the added cost. Mining will become vastly more expensive; and with the rise in the cost of mining there must follow a corresponding rise in the price of coal, iron, and other minerals. The railways, which have as yet failed entirely to develop a satisfactory substitute for the wooden tie (and must, in the opinion of their best engineers, continue to fail), will be profoundly affected, and the cost of transportation will suffer a corresponding increase. Water power for lighting, manu-

[16]

facturing, and transportation, and the movement of freight and passengers by inland waterways, will be affected still more directly than the steam railways. The cultivation of the soil, with or without irrigation, will be hampered by the increased cost of agricultural tools, fencing, and the wood needed for other purposes about the farm. Irrigated agriculture will suffer most of all, for the destruction of the forests means the loss of the waters as surely as night follows day. With the rise in the cost of producing food, the cost of food itself will rise. Commerce in general will necessarily be affected by the difficulties of the primary industries upon which it depends. In a word, when the forests fail, the daily life of the average citizen will inevitably feel the pinch on every side. And the forests have already begun to fail, as the direct result of the suicidal policy of forest destruction which the people of the United States have allowed themselves to pursue.

[17]

It is true that about twenty per cent. of the less valuable timber land in the United States remains in the possession of the people in the National Forests, and that it is being cared for and conserved to supply the needs of the present and to mitigate the suffering of the near future. But it needs no argument to prove that this comparatively small area will be insufficient to meet the demand which is now exhausting an area four times as great, or to prevent the suffering I have described. Measures of greater vigor are imperatively required.

The conception that water is, on the whole, the most important natural resource has gained firm hold in the irrigated West, and is making rapid progress in the humid East. Water, not land, is the primary value in the Western country, and its conservation and use to irrigate land is the first condition of prosperity. The use of our streams for irrigation and for domestic and manufacturing uses is comparatively

[18]

well developed. Their use for power is less developed, while their use for transportation has only begun. The conservation of the inland waterways of the United States for these great purposes constitutes, perhaps, the largest single task which now confronts the Nation. The maintenance and increase of agriculture, the supply of clear water for domestic and manufacturing uses, the development of electrical power, transportation, and lighting, and the creation of a system of inland transportation by water whereby to regulate freight-rates by rail and to move the bulkier commodities cheaply from place to place, is a task upon the successful accomplishment of which the future of the Nation depends in a peculiar degree.

We are accustomed, and rightly accustomed, to take pride in the vigorous and healthful growth of the United States, and in its vast promise for the future. Yet we are making no preparation to realize what we so easily foresee and glibly predict.

THE FIGHT FOR CONSERVATION

The vast possibilities of our great future will become realities only if we make ourselves, in a sense, responsible for that future. The planned and orderly development and conservation of our natural resources is the first duty of the United States. It is the only form of insurance that will certainly protect us against the disasters that lack of foresight has in the past repeatedly brought down on nations since passed away.

CHAPTER II

HOME-BUILDING FOR THE NATION

THE most valuable citizen of this or any other country is the man who owns the land from which he makes his living. No other man has such a stake in the country. No other man lends such steadiness and stability to our national life. Therefore no other question concerns us more intimately than the question of homes. Permanent homes for ourselves, our children, and our Nation — this is a central problem. The policy of national irrigation is of value to the United States in very many ways, but the greatest of all is this, that national irrigation multiplies the men who own the land from which they make their living. The old saying, "Who ever heard of a man shouldering his gun to fight for his boarding house?" reflects this great truth, that no man is so ready to defend his country, not only with arms, but with his vote and his contribution to public opinion, as the man with a permanent stake in it, as the man who owns the land from which he makes his living.

Our country began as a nation of farmers. During the periods that gave it its character, when our independence was won and when our Union was preserved, we were pre-eminently a nation of farmers. We can not, and we ought not, to continue exclusively, or even chiefly, an agricultural country, because one man can raise food enough for many. But the farmer who owns his land is still the backbone of this Nation; and one of the things we want most is more of him. The man on the farm is valuable to the Nation, like any other citizen, just in proportion to his intelligence, character, ability, and patriotism;

but, unlike other citizens, also in proportion to his attachment to the soil. That is the principal spring of his steadiness, his sanity, his simplicity and directness, and many of his other desirable qualities. He is the first of home-makers.

The nation that will lead the world will be a Nation of Homes. The object of the great Conservation movement is just this, to make our country a permanent and prosperous home for ourselves and for our children, and for our children's children, and it is a task that is worth the best thought and effort of any and all of us.

To achieve this or any other great result, straight thinking and strong action are necessary, and the straight thinking comes first. To make this country what we need to have it, we must think clearly and directly about our problems, and above all we must understand what the real problems are. The great things are few and simple, but they are too often hidden by false issues,

[23]

and conventional, unreal thinking. The easiest way to hide a real issue always has been, and always will be, to replace it with a false one.

The first thing we need in this country, as President Roosevelt so well set forth in a great message which told what he had been trying to do for the American people, is equality of opportunity for every citizen. No man should have less, and no man ought to ask for any more. Equality of opportunity is the real object of our laws and institutions. Our institutions and our laws are not valuable in themselves. They are valuable only because they secure equality of opportunity for happiness and welfare to our citizens. An institution or a law is a means, not an end, a means to be used for the public good, to be modified for the public good, and to be interpreted for the public good. One of the great reasons why President Roosevelt's administration was of such enormous value to the

[24]

plain American was that he understood what St. Paul meant when he said: "The letter killeth, but the spirit giveth life." To follow blindly the letter of the law, or the form of an institution, without intelligent regard both for its spirit and for the public welfare, is very nearly as dangerous as to disregard the law altogether. What we need is the use of the law for the public good, and the construction of it for the public welfare.

It goes without saying that the law is supreme and must be obeyed. Civilization rests on obedience to law. But the law is not absolute. It requires to be construed. Rigid construction of the law works, and must work, in the vast majority of cases, for the benefit of the men who can hire the best lawyers and who have the sources of influence in lawmaking at their command. Strict construction necessarily favors the great interests as against the people, and in the long run can not do otherwise. Wise execution of the law must consider what the law ought to accomplish for the general good. The great oppressive trusts exist because of subservient law-makers and adroit legal constructions. Here is the central stronghold of the money power in the everlasting conflict of the few to grab, and the many to keep or win the rights they were born with. Legal technicalities seldom help the people. The people, not the law, should have the benefit of every doubt.

Equality of opportunity, a square deal for every man, the protection of the citizen against the great concentrations of capital, the intelligent use of laws and institutions for the public good, and the conservation of our natural resources, not for the trusts, but for the people; these are real issues and real problems. Upon such things as these the perpetuity of this country as a nation of homes really depends. We are coming to see that the simple things are the things to work for. More than that, we are coming to see that the plain American citizen is the man to work for. The imagination is staggered by the magnitude of the prize for which we work. If we succeed, there will exist upon this continent a sane, strong people, living through the centuries in a land subdued and controlled for the service of the people, its rightful masters, owned by the many and not by the few. If we fail, the great interests, increasing their control of our natural resources, will thereby control the country more and more, and the rights of the people will fade into the privileges of concentrated wealth.

There could be no better illustration of the eager, rapid, unwearied absorption by capital of the rights which belong to all the people than the water-power trust, perhaps not yet formed but in process of formation. This statement is true, but not unchallenged. We are met at every turn by the indignant denial of the water-power interests. They tell us that there is no community of interest among them, and yet they appear by their paid attorneys, year after year, at irrigation and other congresses, asking for help to remove the few remaining obstacles to their perpetual and complete absorption of the remaining water-powers. They tell us it has no significance that there is hardly a bank in some sections of the country that is not an agency for water-power capital, or that the General Electric Company interests are acquiring great groups of water-powers in various parts of the United States, and dominating the power market in the region of each group. And whoever dominates power, dominates all industry.

Have you ever seen a few drops of oil scattered on the water spreading until they formed a continuous film, which put an end at once to all agitation of the surface? The time for us to agitate this question is now, before the separate circles of centralized control spread into the uniform, unbroken, Nation-wide covering of a single gigantic trust. There will be little chance for mere agitation after that. No man at all familiar with the situation can doubt that the time for effective protest is very short. If we do not use it to protect ourselves now, we may be very sure that the trust will give hereafter small consideration to the welfare of the average citizen when in conflict with its own.

The man who really counts is the plain American citizen. This is the man for whom the Roosevelt policies were created, and his welfare is the end to which the Roosevelt policies lead.

I stand for the Roosevelt policies because they set the common good of all of us above the private gain of some of us; because they recognize the livelihood of the small man as more important to the Nation than the profit of the big man; because they oppose all useless waste at present at the cost of robbing the future; because they demand the complete, sane, and orderly development of all our natural resources; because they insist upon equality of opportunity and denounce monopoly and special privilege; because, discarding false issues, they deal directly with the vital questions that really make a difference with the welfare of us all; and, most of all, because in them the plain American always and everywhere holds the first place. And I propose to stand for them while I have the strength to stand for anything.

CHAPTER III

BETTER TIMES ON THE FARM

EVER since I came to have first-hand knowledge of irrigation, I have been impressed with the peculiar advantages which surround the irrigation rancher. The high productiveness of irrigated land, resulting in smaller farm units and denser settlement, as well as the efficiency and alertness of the irrigator, have combined to give the irrigated regions very high rank among the most progressive farming communities of the world. Such rural communities as those of the irrigated West are useful examples for the consideration of regions in which life is more isolated, has less of the benefits of coöperation, and generally has lacked the stimulus found in irrigation farming.

The object of education in general is to produce in the boy or girl, and so in the man or woman, three results: first, a sound, useful, and usable body; second, a flexible, well-equipped, and well-organized mind; alert to gain interest and assistance from contact with nature and coöperation with other minds; and third, a wise and true and valiant spirit, able to gather to itself the higher things that best make life worth while. The use and growth of these three things, body, mind, and spirit, must all be found in any effective system of education.

The same three-fold activity is equally necessary in a group of individuals. Take for example the merchants of a town, who have established a Chamber of Commerce or Board of Trade. They have three objects: first, sound and profitable business; second, organized coöperation with each other to their mutual advantage, as in settling disputes, securing satisfactory rates from railroads, and inducing new industries to settle amongst them; and third, to make their town more beautiful, more healthful, and generally a better place to live in. Take a labor union as another example, and you will find the same three-fold purpose. A good union admits only good workmen to membership in its sound body; the members get from the Union the advantages of organized coöperation in selling their labor to the best advantage; and in addition they enjoy certain special advantages often of overwhelming importance.

The practical value of organization and coöperation is obvious, and they are being utilized very widely in nearly every branch of our national life. But what is the case with the farmer? The farmers are the only great body of our people who remain in large part substantially unorganized. The merchants are organized, the wage-workers are organized, the railroads are organized. The men with whom the farmer competes are organized to get the best results

[33]

for themselves in their dealings with him. The farmer is engaged, usually without the assistance of organization, in competing with these organizations of other groups of citizens. Thus the farmer, the man on whose product we all live, too often contends almost single-handed against his highly organized competitors.

How have the agricultural schools and colleges and the Departments of Agriculture of State and Nation met this situation? Largely by the assertion, in word or in act, that there is only one thing to be done for the farmer. So far as his personal education is concerned, they have tried to give him a sound body, a trained mind, and a wise and valiant spirit. But so far as his calling is concerned, they have stopped with the body. They have said in effect: We will help the farmer to grow better crops, but we will take no thought of how he can get the best returns for the crops he grows, or of how he can utilize those returns so as

[34]

to make them yield him the best and happiest life.

It is not wise to stop the education of a boy or a girl with the body, and to neglect the mind and the spirit. But we have done the equivalent of that in dealing with farm life. Along the line of better crops we have done more for the farmer, and have done it more effectively, than any other Nation. But we have done little, and far less than many other Nations, for better business and better living on the farm. Hereafter we shall need in State and Nation not only the work of Departments of Agriculture such as we have now, but we shall need to have added to their functions such duties as will make them departments of rural business and rural life as well. Our Departments of Agriculture should cover the whole field of the farmer's life. It is not enough to touch only one of the three great country problems, even though that is the first in time and perhaps in importance.

[35]

Of course we all realize that the growing of crops is the great foundation on which the well-being not only of the farmer but of the whole Nation must depend. First of all we must have food. But after that has been achieved, is there nothing more to be done? It seems to me clear that farmers have as much to gain from good organization as merchants, plumbers, carpenters, or any of the other trades and businesses of the United States. After we have secured better crops, the next logical and inevitable step is to secure better business organization on the farm, so that each farmer shall get from what he grows the best possible return.

Consider what has been accomplished in Ireland through agricultural coöperation The Irish have discovered that it is not good for the farmer to work alone. Since 1894 they have been organizing agricultural societies to give the farmer a chance to sell at the right time and at the right price. The result is impressive. In Ireland the coöperative creameries produce about half the butter exported. There are 40,000 farmers in the societies for coöperative selling, which, as we know in this country, means better prices. There are about 300 agricultural credit societies with a membership of 15,000 and a capital of more than $200,000. In a word, in Ireland, which we have been apt to consider as far behind us in all that relates to agriculture, there are nearly 1,000 agricultural societies with a total membership of 100,000 persons. Since 1894 their total business has been more than $300,000,000.

But, after the farmer has begun to make use of his right to combine for his advantage in selling his products and buying his supplies, is there nothing else he can do? As well might we say that, after the body and the mind of a boy have been trained, he should be deprived of all those associations with his fellows which make life worth living, and to which every child has an inborn right. Life is something more than a matter of business. No man can make his life what it ought to be by living it merely on a business basis. There are things higher than business. What is the reason for the enormous movement from the farms into the cities? Not simply that the business advantages in the city are better, but that the city has more conveniences, more excitement, and more facility for contact with friends and neighbors: in a word, more life. There ought then to be attractiveness in country life such as will make the country boy or girl want to live and work in the country, such that the farmer will understand that there is no more dignified calling than his own, none that makes life better worth living. The social or community life of the country should be put by the farmer—for no one but himself can do it for him—on the same basis as social life in the city, through the country churches and societies, through better roads, country telephones, rural free delivery, parcels post, and whatever else will help. The problem is not merely to get better crops, not merely to dispose of crops better, but in the last analysis to have happier and richer lives of men and women on the farm.

CHAPTER IV

PRINCIPLES OF CONSERVATION

THE principles which the word Conservation has come to embody are not many, and they are exceedingly simple. I have had occasion to say a good many times that no other great movement has ever achieved such progress in so short a time, or made itself felt in so many directions with such vigor and effectiveness, as the movement for the conservation of natural resources.

Forestry made good its position in the United States before the conservation movement was born. As a forester I am glad to believe that conservation began with forestry, and that the principles which govern the Forest Service in particular and forestry in general are also the ideas that control conservation.

The first idea of real foresight in connection with natural resources arose in connection with the forest. From it sprang the movement which gathered impetus until it culminated in the great Convention of Governors at Washington in May, 1908. Then came the second official meeting of the National Conservation movement, December, 1908, in Washington. Afterward came the various gatherings of citizens in convention, come together to express their judgment on what ought to be done, and to contribute, as only such meetings can, to the formation of effective public opinion.

The movement so begun and so prosecuted has gathered immense swing and impetus. In 1907 few knew what Conservation meant. Now it has become a household word. While at first Conservation was supposed to apply only to forests, we see now that its sweep extends even beyond the natural resources.

The principles which govern the conservation movement, like all great and effective things, are simple and easily understood. Yet it is often hard to make the simple, easy, and direct facts about a movement of this kind known to the people generally.

The first great fact about conservation is that it stands for development. There has been a fundamental misconception that conservation means nothing but the husbanding of resources for future generations. There could be no more serious mistake. Conservation does mean provision for the future, but it means also and first of all the recognition of the right of the present generation to the fullest necessary use of all the resources with which this country is so abundantly blessed. Conservation demands the welfare of this generation first, and afterward the welfare of the generations to follow.

[42]

THE FIGHT FOR CONSERVATION

The first principle of conservation is development, the use of the natural resources now existing on this continent for the benefit of the people who live here now. There may be just as much waste in neglecting the development and use of certain natural resources as there is in their destruction. We have a limited supply of coal, and only a limited supply. Whether it is to last for a hundred or a hundred and fifty or a thousand years, the coal is limited in amount, unless through geological changes which we shall not live to see, there will never be any more of it than there is now. But coal is in a sense the vital essence of our civilization. If it can be preserved, if the life of the mines can be extended, if by preventing waste there can be more coal left in this country after we of this generation have made every needed use of this source of power, then we shall have deserved well of our descendants.

Conservation stands emphatically for the

development and use of water-power now, without delay. It stands for the immediate construction of navigable waterways under a broad and comprehensive plan as assistants to the railroads. More coal and more iron are required to move a ton of freight by rail than by water, three to one. In every case and in every direction the conservation movement has development for its first principle, and at the very beginning of its work. The development of our natural resources and the fullest use of them for the present generation is the first duty of this generation. So much for development.

In the second place conservation stands for the prevention of waste. There has come gradually in this country an understanding that waste is not a good thing and that the attack on waste is an industrial necessity. I recall very well indeed how, in the early days of forest fires, they were considered simply and solely as acts of God, against which any opposition was hopeless and any attempt to control them not merely hopeless but childish. It was assumed that they came in the natural order of things, as inevitably as the seasons or the rising and setting of the sun. To-day we understand that forest fires are wholly within the control of men. So we are coming in like manner to understand that the prevention of waste in all other directions is a simple matter of good business. The first duty of the human race is to control the earth it lives upon.

We are in a position more and more completely to say how far the waste and destruction of natural resources are to be allowed to go on and where they are to stop. It is curious that the effort to stop waste, like the effort to stop forest fires, has often been considered as a matter controlled wholly by economic law. I think there could be no greater mistake. Forest fires were allowed to burn long after the people had means to stop them. The idea that men were helpless in the face

of them held long after the time had passed
when the means of control were fully within
our reach. It was the old story that "as
a man thinketh, so is he"; we came to see
that we could stop forest fires, and we found
that the means had long •been at hand.
When at length we came to see that the control of logging in certain directions was
profitable, we found it had long been possible.
In all these matters of waste of natural
resources, the education of the people to
understand that they can stop the leakage
comes before the actual stopping and after
the means of stopping it have long been
ready at our hands.

In addition to the principles of development and preservation of our resources
there is a third principle. It is this: The
natural resources must be developed and
preserved for the benefit of the many, and
not merely for the profit of a few. We are
coming to understand in this country that
public action for public benefit has a very

much wider field to cover and a much larger part to play than was the case when there were resources enough for every one, and before certain constitutional provisions had given so tremendously strong a position to vested rights and property in general.

A few years ago President Hadley, of Yale, wrote an article which has not attracted the attention it should. The point of it was that by reason of the XIVth amendment to the Constitution, property rights in the United States occupy a stronger position than in any other country in the civilized world. It becomes then a matter of multiplied importance, since property rights once granted are so strongly entrenched, to see that they shall be so granted that the people shall get their fair share of the benefit which comes from the development of the resources which belong to us all. The time to do that is now. By so doing we shall avoid the difficulties and conflicts which will surely arise if we allow

vested rights to accrue outside the possibility
of governmental and popular control.

The conservation idea covers a wider
range than the field of natural resources
alone. Conservation means the greatest
good to the greatest number for the longest
time. One of its great contributions is
just this, that it has added to the worn and
well-known phrase, "the greatest good to
the greatest number," the additional words
"for the longest time," thus recognizing
that this nation of ours must be made to
endure as the best possible home for all its
people.

Conservation advocates the use of foresight, prudence, thrift, and intelligence in
dealing with public matters, for the same
reasons and in the same way that we each
use foresight, prudence, thrift, and intelligence in dealing with our own private affairs.
It proclaims the right and duty of the people
to act for the benefit of the people. Conservation demands the application of com-

[48]

mon-sense to the common problems for the common good.

The principles of conservation thus described — development, preservation, the common good — have a general application which is growing rapidly wider. The development of resources and the prevention of waste and loss, the protection of the public interests, by foresight, prudence, and the ordinary business and home-making virtues, all these apply to other things as well as to the natural resources. There is, in fact, no interest of the people to which the principles of conservation do not apply.

The conservation point of view is valuable in the education of our people as well as in forestry; it applies to the body politic as well as to the earth and its minerals. A municipal franchise is as properly within its sphere as a franchise for water-power. The same point of view governs in both. It applies as much to the subject of good roads as to waterways, and the training of our people in citizenship is as germane to it as the productiveness of the earth. The application of common-sense to any problem for the Nation's good will lead directly to national efficiency wherever applied. In other words, and that is the burden of the message, we are coming to see the logical and inevitable outcome that these principles, which arose in forestry and have their bloom in the conservation of natural resources, will have their fruit in the increase and promotion of national efficiency along other lines of national life.

The outgrowth of conservation, the inevitable result, is national efficiency. In the great commercial struggle between nations which is eventually to determine the welfare of all, national efficiency will be the deciding factor. So from every point of view conservation is a good thing for the American people.

The National Forest Service, one of the chief agencies of the conservation move-

THE FIGHT FOR CONSERVATION

ment, is trying to be useful to the people of this nation. The Service recognizes, and recognizes it more and more strongly all the time, that whatever it has done or is doing has just one object, and that object is the welfare of the plain American citizen. Unless the Forest Service has served the people, and is able to contribute to their welfare it has failed in its work and should be abolished. But just so far as by coöperation, by intelligence, by attention to the work laid upon it, it contributes to the welfare of our citizens, it is a good thing and should be allowed to go on with its work.

The Natural Forests are in the West. Headquarters of the Service have been established throughout the Western country, because its work cannot be done effectively and properly without the closest contact and the most hearty coöperation with the Western people. It is the duty of the Forest Service to see to it that the timber, water-powers, mines, and every other resource of the forests is used for the benefit of the people who live in the neighborhood or who may have a share in the welfare of each locality. It is equally its duty to coöperate with all our people in every section of our land to conserve a fundamental resource, without which this Nation cannot prosper.

CHAPTER V

WATERWAYS

THE connection between forests and rivers is like that between father and son. No forests, no rivers. So a forester may not be wholly beyond his depth when he talks about streams. The conquest of our rivers is one of the largest commercial questions now before us.

The commercial consequences of river development are incalculable. Its results cannot be measured by the yard-stick of present commercial needs. River improvement means better conditions of transportation than we have now, but it means development too. We cannot see this problem clearly and see it whole in the light of the past alone.

The actual problems of river development

[53]

✳ are not less worthy of our best attention than their commercial results. Every river is a unit from its source to its mouth. If it is to be given its highest usefulness to all the people, and serve them for all the uses they can make of it, it must be developed with ✱ that idea clearly in mind. To develop a river for navigation alone, or power alone, or irrigation alone, is often like using a sheep for mutton, or a steer for beef, and throwing away the leather and the wool. A river is a unit, but its uses are many, and with our present knowledge there can be no excuse for sacrificing one use to another if both can be subserved.

A progressive plan for the development of our waterways is essential. Pending the completion of that plan, which should neither be weakened by excessive haste nor drowned in excessive deliberation, work should proceed at once on some of the greater projects which we know already will be essential under any plan that may be devised. First

and foremost of these by unanimous consent
is the improvement of the Mississippi River.
A comprehensive and progressive plan of
the kind we need can be made in one way
only, and that is by a commission of the
best men in the United States appointed
directly by the President of the United
States.

Such a plan must consider every use to
which our rivers can be put, and every
means available for their control. It must
deal with such great questions as the rela-
tion of the States and the Nation in the
construction and control of the work, and
with terminals and the coördination of
rail and river transportation. The engi-
neering difficulties may be larger than any
we have yet solved. The adjustment of
opposite demands between conflicting interests and localities, and other questions of
large reach and often of great legal com-
plexity will tax the powers of the best men
we have. No part of the work will require

greater temperance, wisdom, and foresight than certain questions of policy and law.

I have observed in the course of some experience that difficulties originating with the law are peculiarly apt to foster misconceptions. It happens that the Forest Service has recently supplied a typical example.

Certain men and certain papers have said that the Forest Service has gone beyond the law in carrying out its work. This assertion has been repeated so persistently that there is danger that it may be believed. The friends of conservation must not be led to think that before the Forest Service can proceed legally with its present work all the hazards and compromises of new legislation must be faced.

Fortunately, the charge of illegal action is absolutely false. The Forest Service has had ample legal authority for everything it has done. Not once since it was created has any charge of illegality, despite the most searching investigation and the bitterest at-

tack, ever led to reversal or reproof by either
House of Congress or by any Congressional
Committee. Since the creation of the Forest Service the expenditure of nearly
$15,000,000 has passed successfully the
scrutiny of the Treasury of the United
States. Most significant of all, not once
has the Forest Service been defeated as to
any vital legal principle underlying its
work in any Court or administrative tribunal
of last resort. Thus those who make the
law and those who interpret it seem to agree
that the work has been legal.

But it is not enough to say that the Forest
Service has kept within the law. Other
qualifications go to make efficiency in a
Government bureau. A bureau may keep
within the law and yet fail to get results.

When action is needed for the public good
there are two opposite points of view regarding the duty of an administrative officer in
enforcing the law. One point of view asks,
"Is there any express and specific law

authorizing or directing such action?" and, having thus sought and found none, nothing is done. The other asks, "Is there any justification in law for doing this desirable thing?" and, having thus sought and found a legal justification, what the public good demands is done. I hold it to be the first duty of a public officer to obey the law. But I hold it to be his second duty, and a close second, to do everything the law will let him do for the public good, and not merely what the law compels or directs him to do.

It is the right as well as the duty of a public officer to be zealous in the public service. That is why the public service is worth while. To every public officer the law should be, not a goad to drive him to his duty, but a tool to help him in his work. And I maintain that it is likewise his right and duty to seek by every proper means from the legal authorities set over him such interpretations of the law as will best help him to serve his country.

THE FIGHT FOR CONSERVATION

Let the public officer take every lawful chance to use the law for the public good. The better use he makes of it the better public servant he becomes. One man with a jack-knife will build a ladder. Another with a full tool-chest cannot make a footstool. The man with the jack-knife will often reach the higher level. I am for the man with the jack-knife. I believe in the man who does all he can and the best he can, with the means at his command. That is precisely what the Forest Service has been trying to do with the money and law Congress has placed in its hands.

Every public officer responsible for any part of the conservation of natural resources is a trustee of the public property. If conservation is vital to the welfare of this Nation now and hereafter, as President Roosevelt so wisely declared, then few positions of public trust are so important, and few opportunities for constructive work so large. Such officers are concerned with the greatest issues which have come before this Nation since the Civil War. They may hope to serve the Nation as few men ever can. Their care for our forests, waters, lands, and minerals is often the only thing that stands between the public good and the something-for-nothing men, who, like the daughters of the horse-leech, are forever crying, "Give, Give." The intelligence, initiative, and steadfastness that can withstand the unrelenting pressure of the special interests are worth having, and the Forest Service has given proof of all three. But the counter-pressure from the people in their own interest is needed far more often than it is supplied.

The public welfare cannot be subserved merely by walking blindly in the old ruts. Times change, and the public needs change with them. The man who would serve the public to the level of its needs must look ahead, and one of his most difficult problems will be to make old tools answer new uses—uses some of which, at least, were never imagined when the tools were made. That is one reason why constructive foresight is one of the great constant needs of every growing nation.

The Forest Service proposes to use the tools — obey the law — made by the representatives of the people. But the law cannot give specific directions in advance to meet every need and detail of administration. The law cannot make brains nor supply conscience. Therefore, the Forest Service proposes also to serve the people by the intelligent and purposeful use of the law and every lawful means at its command for the public good. And for that intention it makes no apology.

Fortunately for the Forest Service, the point of view which it worked out for itself under the pressure of its responsibilities was found to be that of the Supreme Court. In the case of the U. S. vs. Macdaniel (7 Pet., 13-14), involving the administrative powers of the head of a Department,

[61]

the Supreme Court of the United States
said:

"He is limited in the exercise of his
powers by the law; but it does not
follow that he must show statutory
provision for everything he does. No
government could be administered on
such principles. To attempt to regulate, by law, the minute movements
of every part of the complicated machinery of government, would evince a
most unpardonable ignorance on the
subject. Whilst the great outlines of
its movements may be marked out,
and limitations imposed on the exercise of its powers, there are numberless
things which must be done, that can
neither be anticipated nor defined, and
which are essential to the proper action
of the government."

Congress has given to the Secretary of
Agriculture, acting through the Forest Serv-

ice, the specific task of administering the National Forests, with full power to perform it, and, has provided that he "may make such rules and regulations and establish such service as will ensure the objects of said reservations, namely, to regulate their occupancy and use and to preserve the forests thereon from destruction." Every exercise of the powers granted to the Secretary of Agriculture by statute has been in accordance with the principles laid down by Chief Justice Marshall ninety years ago in the case of McCulloch vs. Maryland (4 Wheat., 421), when he said as to powers delegated by the Federal Constitution to Congress:

> "Let the end be legitimate, let it be within the scope of the Constitution, and all means which are appropriate, which are plainly adapted to that end, which are not prohibited, but consist with the letter and spirit of the Constitution, are constitutional."

[63]

After the transfer of the National Forests from the Interior Department to the Forest Service in 1905, some things were done that had never been done before, such as initiating Government control over water-power monopoly in the National Forests, giving preference to the public over commercial corporations in the use of the Forests, and trying to help the small man make a living rather than the big man make a profit (but always with the effort to be just to both). Always and everywhere we have set the public welfare above the advantage of the special interests.

Because it did these things the Forest Service has made enemies, of some of whom it is justly proud. It has been easy for these enemies to raise the cry of illegality, novelty, and excess of zeal. But in every instance the Service has been fortified either by express statutes, or by decisions of the Supreme Court and other courts, of the Secretary of the Interior, of the Comptroller, or the Attor-

[64]

ney-General, or by general principles of law which are beyond dispute. If there is novelty, it consists simply in the way these statutes, decisions, and principles have been used to protect the public. The law officers of the Forest Service have had the Nation for their client, and they are proud to work as zealously for the public as they would in private practice for a fee.

So I think the ghost of illegality in the Forest Service may fairly be laid at rest. But it is not the only one which is clouding the issues of conservation in the public mind. Another misconception is that the friends of conservation are trying to prevent the development of water power by private capital. Nothing could be farther from the truth. The friends of conservation were the first to call public attention to the enormous saving to the Nation which follows the substitution of the power of falling water, which is constantly renewed, for our coal, which can never be renewed.

[65]

They favor development by private capital and not by the Government, but they also favor attaching such reasonable conditions to the right to develop as will protect the public and control water-power monopoly in the public interest, while at the same time giving to enterprising capital its just and full reward. They believe that to grant rights to water power in perpetuity is a wrongful mortgage of the welfare of our descendants, and to grant them without insisting on some return for value received is to rob ourselves.

I believe in dividends for the people as well as taxes. Fifty years is long enough for the certainty of profitable investment in water power, and to fix on the amount of return that will be fair to the public and the corporation is not impossible. What city does not regret some ill-considered franchise? And why should not the Nation profit by the experience of its citizens?

There is no reason why the water-power

interests should be given the people's property freely and forever except that they would like to have it that way. I suspect that the mere wishes of the special interests, although they have been the mainspring of much public action for many years, have begun to lose their compelling power. A good way to begin to regulate corporations would be to stop them from regulating us.

The sober fact is that here is the imminent battle-ground in the endless contest for the rights of the people. Nothing that can be said or done will suffice to postpone longer the active phases of this fight; and that is why I attach so great importance to the attitude of administrative officers in protecting the public welfare in the enforcement of the law.

From time to time a few strong leaders have tried to unite the people in the fight of the many for the equal opportunities to which they are entitled. But the people have only

[67]

just begun to take this fight in earnest. They have not realized until recently the vital importance and far-reaching consequences of their own passive position.

Now that the fight is passing into an acute stage it is easily seen that the special interests have used the period of public indifference to manœuvre themselves into a position of exceeding strength. In the first place, the Constitutional position of property in the United States is stronger than in any other nation. In the second place, it is well understood that the influence of the corporations in our law-making bodies is usually excessive, not seldom to the point of defeating the will of the people steadily and with ease. In the third place, cases are not unknown in which the special interests, not satisfied with making the laws, have assumed also to interpret them, through that worst of evils in the body politic, an unjust judge.

When an interest or an enemy is entrenched

in a position rendered impregnable against an expected mode of attack, there is but one remedy, to shift the ground and follow lines against which no preparation has been made. Fortunately for us, the special interests, with a blindness which naturally follows from their wholly commercialized point of view, have failed to see the essential fact in this great conflict. They do not understand that this is far more than an economic question, that in its essence and in every essential characteristic it is a moral question.

The present economic order, with its face turned away from equality of opportunity, involves a bitter moral wrong, which must be corrected for moral reasons and along moral lines. It must be corrected with justness and firmness, but not bitterly, for that would be to lower the Nation to the moral level of the evil which we have set ourselves to fight.

This is the doctrine of the Square Deal.

[69]

It contains the germ of industrial liberty. Its partisans are the many, its opponents are the few. I am firm in the faith that the great majority of our people are Square Dealers.

CHAPTER VI

BUSINESS

THE business of the people of the United States, performed by the Government of the United States, is a vast and a most important one; it is the house-keeping of the American Nation. As a business proposition it does not attract anything like the attention that it ought. Unfortunately we have come into the habit of considering the Government of the United States as a political organization rather than as a business organization.

Now this question, which the Governors of the States and the representatives of great interests were called to Washington to consider in 1908, is fundamentally a business question, and it is along business lines that it must be considered and solved, if the problem is to be solved at all. Manufacturers are dealing with the necessity for producing a definite output as a result of definite expenditure and definite effort. The Government of the United States is doing exactly the same thing. The manufacturer's product can be measured in dollars and cents. The product of the Government of the United States can be measured partly in dollars and cents, but far more importantly in the welfare and contentment and happiness of the people over which it is called upon to preside.

The keynote of that Conservation Conference in Washington was forethought and foresight. The keynote of success in any line of life, or one of the great keynotes, must be forethought and foresight. If we, as a Nation, are to continue the wonderful growth we have had, it is forethought and foresight which must give us the capacity to go on as we have been going. I dwell on this because it seems to me to be one of the most curious of all things in the history of the United States to-day that we should have grasped this principle so tremendously and so vigorously in our daily lives, in the conduct of our own business, and yet have failed so completely to make the obvious application in the things which concern the Nation.

It is curiously true that great aggregations of individuals and organized bodies are apt to be less far-sighted, less moral, less intelligent along certain lines than the individual citizen; or at least that their standards are lower; a principle which is illustrated by the fact that we have got over settling disputes between individuals by the strong hand, but not yet between nations.

So we have allowed ourselves as a Nation, in the flush of the tremendous progress that we have made, to fail to look at the end from the beginning and to put ourselves in a position where the normal operation of natural laws threatens to bring us to a halt in a way which will make every man, woman, and child in the Nation feel the pinch when it comes.

No man may rightly fail to take a great pride in what has been accomplished by means of the destruction of our natural resources so far as it has gone. It is a paradoxical statement, perhaps, but nevertheless true, because out of this attack on what nature has given we have won a kind of prosperity and a kind of civilization and a kind of man that are new in the world. For example, nothing like the rapidity of the destruction of American forests has ever been known in forest history, and nothing like the efficiency and vigor and inventiveness of the American lumberman has ever been developed by any attack on any forests elsewhere. Probably the most effective tool that the human mind and hand have ever made is the American axe. So the American business man has grasped his opportunities

[74]

and used them and developed them and in-
vented about them, thought them into lines
of success, and thus has developed into a new
business man, with a vigor and effectiveness
and a cutting-edge that has never been
equalled anywhere else. We have gained
out of the vast destruction of our natural
resources a degree of vigor and power and
efficiency of which every man of us ought to
be proud.

Now that is done. We have accomplished
these big things. What is the next step?
Shall we go on in the same lines to the
certain destruction of the prosperity which
we have created, or shall we take the obvious
lesson of all human history, turn our backs
on the uncivilized point of view, and adopt
toward our natural resources the average
prudence and average foresight and average
care that we long ago adopted as a rule of
our daily life?

The conservation movement is calling
the attention of the American people to the

fact that they are trustees. The fact seems to me so plain as to require only a statement of it, to carry conviction. Can we reasonably fail to recognize the obligation which rests upon us in this matter? And, if we do fail to recognize it, can we reasonably expect even a fairly good reputation at the hands of our descendants?

Business prudence and business common-sense indicate as strongly as anything can the absolute necessity of a change in point of view on the part of the people of the United States regarding their natural resources. The way we have been handling them is not good business. Purely on the side of dollars and cents, it is not good business to kill the goose that lays the golden egg, to burn up half our forests, to waste our coal, and to remove from under the feet of those who are coming after us the opportunity for equal happiness with ourselves. The thing we ought to leave to them is not merely an opportunity for equal happiness

[76]

and equal prosperity, but for a vastly increased fund of both.

Conservation is not merely a question of business, but a question of a vastly higher duty. In dealing with our natural resources we have come to a place at last where every consideration of patriotism, every consideration of love of country, of gratitude for things that the land and the institutions of this Nation have given us, call upon us for a return. If we owe anything to the United States, if this country has been good to us, if it has given us our prosperity, our education, and our chance of happiness, then there is a duty resting upon us. That duty is to see, so far as in us lies, that those who are coming after us shall have the same opportunity for happiness we have had ourselves. Apart from any business consideration, apart from the question of the immediate dollar, this problem of the future wealth and happiness and prosperity of the people of the United States has a right to our attention.

[77]

THE FIGHT FOR CONSERVATION

It rises far above all matters of temporary individual business advantage, and becomes a great question of national preservation. We all have the unquestionable right to a reasonable use of natural resources during our lifetime, we all may use, and should use, the good things that were put here for our use, for in the last analysis this question of conservation is the question of national preservation and national efficiency.

CHAPTER VII

THE MORAL ISSUE

THE central thing for which Conservation stands is to make this country the best possible place to live in, both for us and for our descendants. It stands against the waste of the natural resources which cannot be renewed, such as coal and iron; it stands for the perpetuation of the resources which can be renewed, such as the food-producing soils and the forests; and most of all it stands for an equal opportunity for every American citizen to get his fair share of benefit from these resources, both now and hereafter.

Conservation stands for the same kind of practical common-sense management of this country by the people that every business man stands for in the handling of his own business. It believes in prudence and foresight instead of reckless blindness; — it holds that resources now public property should not become the basis for oppressive private monopoly; and it demands the complete and orderly development of all our resources for the benefit of all the people, instead of the partial exploitation of them for the benefit of a few. It recognizes fully the right of the present generation to use what it needs and all it needs of the natural resources now available, but it recognizes equally our obligation so to use what we need that our descendants shall not be deprived of what they need.

Conservation has much to do with the welfare of the average man of to-day. It proposes to secure a continuous and abundant supply of the necessaries of life, which means a reasonable cost of living and business stability. It advocates fairness in the distribution of the benefits which flow from the natural resources. It will matter very little to the average citizen, when scarcity comes and prices rise, whether he can not get what he needs because there is none left or because he can not afford to pay for it. In both cases the essential fact is that he can not get what he needs. Conservation holds that it is about as important to see that the people in general get the benefit of our natural resources as to see that there shall be natural resources left.

Conservation is the most democratic movement this country has known for a generation. It holds that the people have not only the right, but the duty to control the use of the natural resources, which are the great sources of prosperity. And it regards the absorption of these resources by the special interests, unless their operations are under effective public control, as a moral wrong. Conservation is the application of common-sense to the common problems for the common good, and I believe it stands nearer to the desires, aspirations, and purposes of the average man than any other policy now before the American people.

The danger to the Conservation policies is that the privileges of the few may continue to obstruct the rights of the many, especially in the matter of water power and coal. Congress must decide immediately whether the great coal fields still in public ownership shall remain so, in order that their use may be controlled with due regard to the interest of the consumer, or whether they shall pass into private ownership and be controlled in the monopolistic interest of a few.

Congress must decide also whether immensely valuable rights to the use of water power shall be given away to special interests in perpetuity and without compensation instead of being held and controlled by the public. In most cases actual development of water power can best be done

[82]

by private interests acting under public
control, but it is neither good sense nor
good morals to let these valuable privileges
pass from the public ownership for nothing
and forever. Other conservation matters
doubtless require action, but these two,
the conservation of water power and of
coal, the chief sources of power of the
present and the future, are clearly the most
pressing.

It is of the first importance to prevent
our water powers from passing into private
ownership as they have been doing, because
the greatest source of power we know is
falling water. Furthermore, it is the only
great unfailing source of power. Our coal,
the experts say, is likely to be exhausted
during the next century, our natural gas
and oil in this. Our rivers, if the forests
on the watersheds are properly handled,
will never cease to deliver power. Under
our form of civilization, if a few men ever
succeed in controlling the sources of power,

[83]

they will eventually control all industry as well. If they succeed in controlling all industry, they will necessarily control the country. This country has achieved political freedom; what our people are fighting for now is industrial freedom. And unless we win our industrial liberty, we can not keep our political liberty. I see no reason why we should deliberately keep on helping to fasten the handcuffs of corporate control upon ourselves for all time merely because the few men who would profit by it most have heretofore had the power to compel it.

The essential things that must be done to protect the water powers for the people are few and simple. First, the granting of water powers forever, either on non-navigable or navigable streams, must absolutely stop. It is perfectly clear that one hundred, fifty, or even twenty-five years ago our present industrial conditions and industrial needs were completely beyond

[84]

the imagination of the wisest of our pred-
ecessors. It is just as true that we can
not imagine or foresee the industrial conditions and needs of the future. But we
do know that our descendants should be
left free to meet their own necessities as
they arise. It can not be right, therefore,
for us to grant perpetual rights to the one
great permanent source of power. It is
just as wrong as it is foolish, and just as
needless as it is wrong, to mortgage the
welfare of our children in such a way as
this. Water powers must and should be
developed mainly by private capital and
they must be developed under conditions
which make investment in them profitable
and safe. But neither profit nor safety
requires perpetual rights, as many of the best
water-power men now freely acknowledge.

Second, the men to whom the people
grant the right to use water-power should
pay for what they get. The water-power
sites now in the public hands are enor-

[85]

mously valuable. There is no reason whatever why special interests should be allowed
to use them for profit without making some
direct payment to the people for the valuable
rights derived from the people. This is
important not only for the revenue the Nation
will get. It is at least equally important
as a recognition that the public controls its
own property and has a right to share in
the benefits arising from its development.
There are other ways in which public control of water power must be exercised, but
these two are the most important.

Water power on non-navigable streams
usually results from dropping a little water
a long way. In the mountains water is
dropped many hundreds of feet upon the
turbines which move the dynamos that produce the electric current. Water power on
navigable streams is usually produced by
dropping immense volumes of water a short
distance, as twenty feet, fifteen feet, or
even less. Every stream is a unit from its

source to its mouth, and the people have
the same stake in the control of water power
in one part of it as in another. Under the
Constitution, the United States exercises
direct control over navigable streams. It
exercises control over non-navigable and
source streams only through its ownership
of the lands through which they pass, as
the public domain and National Forests.
It is just as essential for the public welfare
that the people should retain and exercise
control of water-power monopoly on navigable as on non-navigable streams. If the
difficulties are greater, then the danger
that the water powers may pass out of the
people's hands on the lower navigable parts
of the streams is greater than on the upper
non-navigable parts, and it may be harder,
but in no way less necessary, to prevent it.

It must be clear to any man who has
followed the development of the Conservation idea that no other policy now before
the American people is so thoroughly democratic-

cratic in its essence and in its tendencies as the Conservation policy. It asserts that the people have the right and the duty, and that it is their duty no less than their right, to protect themselves against the uncontrolled monopoly of the natural resources which yield the necessaries of life. We are beginning to realize that the Conservation question is a question of right and wrong, as any question must be which may involve the differences between prosperity and poverty, health and sickness, ignorance and education, well-being and misery, to hundreds of thousands of families. Seen from the point of view of human welfare and human progress, questions which begin as purely economic often end as moral issues. Conservation is a moral issue because it involves the rights and the duties of our people — their rights to prosperity and happiness, and their duties to themselves, to their descendants, and to the whole future progress and welfare of this Nation.

[88]

CHAPTER VIII

PUBLIC SPIRIT

VIOLENT crises in the lives of men and nations usually produce their own remedies. They grasp the attention and stir the consciences of men, and usually they evolve leaders and measures to meet their imperious needs. But the great evident crises are by no means the only ones of importance. The quiet turning point, reached and passed often with slight attention and wholly without struggle, is frequently not less decisive. Great decisions are made or great impulses given or withheld in the life of a man or a nation often so quietly that their critical character is seen only in retrospect. It is only the historian who can say just when some unnoticed, yet decisive

[89]

and irrevocable, step was actually accomplished.

The United States has been in the midst of such a period of decision since the Spanish War called into blossom the quiet growth of years, and we are still face to face with questions of the most vital bearing upon our future. The changes now in progress are accompanied by no convulsions, yet the whole character of our civilization is being rapidly crystallized anew as our country takes its inevitable place in the world.

So quietly are the great forces at work that some of our most vital problems have remained almost unrecognized by the public until the last two years. Yet the fact that these decisions are being made is almost appalling in its magnitude, and their indescribable consequence not only to the United States, but to all the nations of the earth, needs to be vividly realized by every one of us, for it is one of the great compelling reasons why the public spirit of young men is needed so urgently and at once. And more specific reasons press upon us from every side.

Recently the attention of our people, thanks largely to President Roosevelt, was focussed upon the presence or absence of the common virtues and the common decencies in public life. The revelation of corruption in politics, in business, and here and there in the public service, is a testimony not of unwonted wickedness in high places, but of unwonted sensitiveness in public opinion, and so far as it goes it is a most hopeful sign; but it does not yet go far enough.

The opportunity to set a new standard in political morality is here now. Public sensitiveness on every subject ebbs and flows and must be taken at the flood if the use of it is to be really effective. Decision made now as to the character of our public life will be valid for many years, for it is but seldom that the question comes so clearly be-

fore us. The war for righteousness is endless, but this is one of the great battles,
and its results will endure.

We are now in the throes of decision on
the whole question of business in politics,
of politics for business purposes, and we
must take our share in determining whether
the object of our political system is to be
unclean money or free men. The present
strong movement to prevent the political
control of public men, law-courts, and legislatures by great commercial enterprises
will either flash in the pan or it will succeed; it will leave either the man or the
dollar in control. The decision will be
made by the young men, and it is not
far ahead.

The question of efficiency in public office
has been brought to the front as never
before in the history of the Nation. As a
whole, our public service is honest, but we
should be able to take honesty for granted.
What we lack is the tradition of high effi-

ciency that makes great enterprises succeed. The national housekeeping, the Government's vast machinery, should be the cleanest, the most effective, and the best in methods and in men, for its touch upon the life of the Nation at every point is constant and vital.

There is no hunger like land hunger, and no object for which men are more ready to use unfair and desperate means than the acquisition of land. Under the influence of this compelling desire, assisted by obsolete land laws warped from their original purpose, we are facing in the public-land States west of the Mississippi the great question whether the Western people are to be predominately a people of tenants under the degrading tyranny of pecuniary and political vassalage, or freeholders and free men; and there is no exaggerating the importance of the decision.

We have been deciding, and the decision is not yet fully made, whether the future shall suffer the long train of ills which everywhere has followed, and must always follow, the abuse of the forest, or whether by protecting the timberlands we shall assure the prosperity of all of the users of the wood, the water, and the forage which our forests supply. Nothing less than the whole agricultural and commercial welfare of the country is in the balance. No other conservation question compares with this in the vital intimacy of its touch on every portion of our national life.

Other great questions only less vital I cannot even refer to, but one of the central ones remains — our whole future is at stake in the education of our young men in politics and public spirit. The greatest work that Theodore Roosevelt did for the United States, the great fact which will give his influence vitality and power long after we shall all have gone to our reward, greater than his great services in bringing peace, in settling strikes, in preaching the crusade of honesty and decency in business and in daily life, is the fact that he changed the attitude of the American people toward conserving the natural resources, and toward public questions and public life. The time was, not long ago, when it was not respectable to be interested in politics. The time is coming, and I do not believe it is far ahead, when it will not be respectable not to be interested in public affairs. Few changes can mean so much.

Among the first duties of every man is to help in bringing the Kingdom of God on earth. The greatest human power for good, the most efficient earthly tool for the future uplifting of the nations, is without question the United States; and the presence or absence of a vital public spirit in the young men of the United States will determine the quality of that great tool and the work that it can do. This is the final object of the best citizenship. Public spirit is the means by which every man can help

[95]

toward this great end. Public spirit is patriotism in action; it is the application of Christianity to the commonwealth; it is effective loyalty to our country, to the brotherhood of man, and to the future. It is the use of a man by himself for the general good.

Public spirit is the one great antidote for all the ills of the Nation, and greatly the Nation needs it now. In a day when the vast increase in wealth tends to reduce all things, moral, intellectual and material, to the measure of the dollar; in a day when we have with us always the man who is working for his own pocket all the time; when the monopolist of land, of opportunity, of power or privilege in any form, is ever in the public eye — it is good to remember that the real leaders are the men who value the right to give themselves more highly than any gain whatsoever.

It is given to few men to serve their country as greatly as President Roosevelt has done, yet vastly smaller services are still tremendously worth while. I question whether there has ever been a time and place (except in violent crises) when the demand for public spirit was greater than now and the results of it more assured. Public spirit is never needed more than in times of prosperity, and it is never more effective. It is the boat which is floating easily and rapidly with the stream that is most in danger of striking the rocks.

The reasons why public opinion may be so effective in the United States are not far to seek. The extreme sensitiveness of our form of government to political control is one of the commonplaces that has real meaning. We seldom realize that ours is actually what it pretends to be — a representative government — and our legislatures are extraordinarily sensitive to what the people, the politically effective people, really want. The Senators and Representatives in Congress do actually and accurately represent the men who send them there, and they respond like lightning to a clear order from the controlling element at home. It is in the power of public spirit to say whether men or money shall control.

If public spirit is in the saddle, the fundamental purpose of all the people, which is good, will govern. If not, the bosses and the great private interests will have their way. Without the backing of the public spirit of good men, even the President himself loses by far the greater portion of his power. For the power to do what we hope to see accomplished, we must look most of all to the public spirit of the young men.

But some one will say that great service is beyond his individual power. I do not believe that great service is beyond the power of any young man. This is not a matter in which obstacles decide. The man for whom all the barriers to success have been broken down is not, as a rule, the man who

[98]

succeeds. On the contrary, conflict is the condition of success. The quality of the man himself decides. The more I study men, which is the daily occupation of every man in affairs, the more firmly I am assured that the great fundamental difference between men, the reason why some fail and some succeed, is not a difference in ability or opportunity, but a difference in vision and in relentless loyalty to ideals — vision to see the great object, and relentless, unwavering, uninterrupted loyalty in its service. What young men determine to do at whatever cost of effort, self-denial, and endurance, provided that their objects are good and within the possibility of attainment, they will surely accomplish in so large a proportion of cases that the failures are negligible. If all that a man has or is, if his death and his daily life, are wholly and relentlessly at the service of his ideal, without hesitancy or reservation, then he will achieve his object. Either by himself or his successors he will achieve it, for he disposes of the greatest power to which humanity can attain. Under such conditions there is no man among us who cannot render high service to our beloved country.

CHAPTER IX

THE CHILDREN

THE success of the conservation movement in the United States depends
in the end on the understanding the
women have of it. No forward step in
this whole campaign has been more
deeply appreciated or more welcomed than
that which the National Society of the
Daughters of the American Revolution and
other organizations of women have taken in
appointing conservation committees.

Patriotism is the key to the success of
any nation, and patriotism first strikes its
roots in the mind of the child. Patriotism
which does not begin in early years may,
though it does not always, fail under the
severest trials. I say "not always," for

[101]

many men and women have proved their patriotic devotion to this country although they were born elsewhere. Yet, as a rule, it must begin with the children. And almost without exception it is the mother who plants patriotism in the mind of the child. It is her duty. The growth of patriotism is first of all in the hands of the women of any nation. In the last analysis it is the mothers of a nation who direct that nation's destiny.

The fundamental task of patriotism is to see to it that the Nation exists and endures in honor, security, and well-being. Fortunately there is no question as to our existing in honor, and little if any as to our continuing to exist in security.

The great fundamental problem which confronts us all now is this: Shall we continue, as a Nation, to exist in well-being? That is the conservation problem.

If we are to have prosperity in this country, it will be because we have an abundance of natural resources available for the citizen.

In other words, as the minds of the children are guided toward the idea of foresight, just to that extent, and probably but little more, will the generations that are coming hereafter be able to carry through the great task of making this Nation what its manifest destiny demands that it shall be.

Women should recognize, if this task is to be carried out, one great truth above all others. That this Nation exists for its people, we all admit; but that the natural resources of the Nation exist not for any small group, not for any individual, but for all the people — in other words, that the natural resources of the Nation belong to all the people — that is a truth the whole meaning of which is just beginning to dawn on us. There is no form of monopoly which exists or ever has existed on any large scale which was not based more or less directly upon the control of natural resources. There is no form of monopoly that has ever existed or can exist which can do harm if the people understand that the natural resources belong to the people of the Nation, and exercise that understanding, as they have the power to do.

It seems to me that of all the movements which have been inaugurated to give power to the conservation idea, the foresight idea, there is none more helpful than that the women of the United States are taking hold of the problem. We must make all the people see that now and in the future the resources are to be developed and employed, yet at the same time guarded and protected against waste — not for small groups of men who will control them for their own purposes, but for all the people through all time.

The question of the conservation of our natural resources is not a simple question, but it requires, and will increasingly require, thinking out along lines directed to the fundamental economic basis upon which this Nation exists. I think it can not be disputed that the natural resources exist for and belong to the people; and I believe that the part of the work which falls to the women (and it is no small part) is to see to it that the children, who will be the men and women of the future, have their share of these resources uncontrolled by monopoly and unspoiled by waste.

What specific things can the women of the Nation do for conservation? The Daughters of the American Revolution have begun admirably in the appointment of a Conservation Committee, and other organizations of women are following their example. Few people realize what women have already done for conservation, and what they may do. Some of the earliest effective forest work that was done in the United States, work which laid the lines that have been followed since, was that of the Pennsylvania Forestry Association, begun and carried through first of all by ladies in Philadelphia. One of the bravest, most

intelligent and most effective fights for forestry that I have known of was that of the
women of Minnesota for the Minnesota
National Forest. It was a superb success, and we have that forest to-day. I
have known of no case of persistent agitation under discouragement finer in a good
many ways than the fight that the women
of California have made to save the great
grove of Calaveras big trees. As a result
the Government has taken possession of
that forest and will preserve it for all future
generations.

Time and again, then, the women have
made it perfectly clear what they can do
in this work. Obviously the first point of
attack is the stopping of waste. Women alone
can bring to the school children the idea
of the wickedness of national waste and the
value of public saving. The issue is a moral
one; and women are the first teachers of
right and wrong. It is a question of seeing
what loyalty to the public welfare demands

of us, and then of caring enough for the public welfare not to set personal advantage first. It is a question of inspiring our future citizens while they are boys and girls with the spirit of true patriotism as against the spirit of rank selfishness, the anti-social spirit of the man who declines to take into account any other interest than his own; whose one aim and ideal is personal success. Women both in public and at home, by letting the men know what they think, and by putting it before the children, can make familiar the idea of conservation, and support it with a convincingness that nobody else can approach.

However important it may be for the lumberman, the miner, the wagon-maker, the railroad man, the house-builder,—for every industry,—that conservation should obtain, when all is said and done, conservation goes back in its directest application to one body in this country, and that is to the children. There is in this country no other movement except possibly the education movement — and that after all is in a sense only another aspect of the conservation question, the seeking to make the most of what we have — so directly aimed to help the children, so conditioned upon the needs of the children, so belonging to the children, as the conservation movement; and it is for that reason more than any other that it has the support of the women of the Nation.

CHAPTER X

AN EQUAL CHANCE

THE American people have evidently made up their minds that our natural resources must be conserved. That is good. but it settles only half the question. For whose benefit shall they be conserved — for the benefit of the many, or for the use and profit of the few? The great conflict now being fought will decide. There is no other question before us that begins to be so important, or that will be so difficult to straddle, as the great question between special interest and equal opportunity, between the privileges of the few and the rights of the many, between government by men for human welfare and government by money for profit, between the men who stand for the Roosevelt policies and the men who stand against them. This is the heart of the conservation problem to-day.

The conservation issue is a moral issue. When a few men get possession of one of the necessaries of life, either through owner-ship of a natural resource or through unfair business methods, and use that control to extort undue profits, as in the recent cases of the Sugar Trust and the beef-packers, they injure the average man without good reason, and they are guilty of a moral wrong. It does not matter whether the undue profit comes through stifling competition by rebates or other crooked devices, through corruption of public officials, or through seizing and monopolizing resources which belong to the people. The result is always the same — a toll levied on the cost of living through special privilege.

The income of the average family in the United States is less than $600 a year. To increase the cost of living to such a family beyond the reasonable profits of legitimate business is wrong. It is not merely a question of a few cents more a day for the necessaries of life, or of a few cents less a day for wages. Far more is at stake — the health or sickness of little babies, the education or ignorance of children, virtue or vice in young daughters, honesty or criminality in young sons, the working power of bread-winners, the integrity of families, the provision for old age — in a word, the welfare and happiness or the misery and degradation of the plain people are involved in the cost of living.

To the special interest an unjust rise in the cost of living means simply higher profit, but to those who pay it, that profit is measured in schooling, warm clothing, a reserve to meet emergencies, a fair chance to make the fight for comfort, decency, and right living.

I believe in our form of government and I believe in the Golden Rule. But we must face the truth that monopoly of the sources of production makes it impossible for vast numbers of men and women to earn a fair living. Right here the conservation question touches the daily life of the great body of our people, who pay the cost of special privilege. And the price is heavy.. That price may be the chance to save the boys from the saloons and the corner gang, and the girls from worse, and to make good citizens of them instead of bad; for an appalling proportion of the tragedies of life spring directly from the lack of a little money. Thousands of daughters of the poor fall into the hands of the white-slave traders because their poverty leaves them without protection. Thousands of families, as the Pittsburg survey has shown us, lead lives of brutalizing overwork in return for the barest living. Is it fair that these thousands of families should have less than they need in order that a few families should have swollen fortunes at their expense? Let him who dares deny that there is wickedness in grinding the faces of the poor, or assert that these are not moral questions which strike the very homes of our people. If these are not moral questions, there are no moral questions.

The people of this country have lost vastly more than they can ever regain by gifts of public property, forever and without charge, to men who gave nothing in return. It is true that we have made superb material progress under this system, but it is not well for us to rejoice too freely in the slices the special interests have given us from the great loaf of the property of all the people.

The people of the United States have been the complacent victims of a system of grab, often perpetrated by men who would have been surprised beyond measure to be accused of wrong-doing, and many of whom in their private lives were model citizens. But they have suffered from a curious moral perversion by which it becomes praiseworthy to do for a corporation things which they would refuse with the loftiest scorn to do for themselves. Fortunately for us all that delusion is passing rapidly away.

President Hadley well said that "the fundamental division of powers in the Constitution of the United States is between voters on the one hand and property-owners on the other." When property gets possession of the voting power also, little is left for the people. That is why the unholy alliance between business and politics is the most dangerous fact in our political life. I believe the American people are tired of that alliance. They are weary of politics for revenue only. It is time to take business out of politics, and keep it out — time for the political activity of this Nation to be aimed squarely at the welfare of all of us, and squarely away from the excessive profits of a few of us.

A man is not bad because he is rich, nor good because he is poor. There is no monopoly of virtue. I hold no brief for the

poor against the rich nor for the wage-
earner against the capitalist. Exceptional
capacity in business, as in any other line
of life, should meet with exceptional reward.
Rich men have served this country greatly.
Washington was a rich man. But it is
very clear that excessive profits from the
control of natural resources, monopolized
by a few, are not worth to this Nation the
tremendous price they cost us.

We have allowed the great corporations
to occupy with their own men the strategic
points in business, in social, and in political
life. It is our fault more than theirs. We
have allowed it when we could have stopped
it. Too often we have seemed to forget
that a man in public life can no more serve
both the special interests and the people
than he can serve God and Mammon.
There is no reason why the American people
should not take into their hands again the
full political power which is theirs by right,
and which they exercised before the special

[115]

interests began to nullify the will of the majority. There are many men who believe, and who will always believe, in the divine right of money to rule. With such men argument, compromise, or conciliation is useless or worse. The only thing to do with them is to fight them and beat them. It has been done, and it can be done again.

It is the honorable distinction of the Forest Service that it has been more constantly, more violently and more bitterly attacked by the representatives of the special interests in recent years than any other Government Bureau. These attacks have increased in violence and bitterness just in proportion as the Service has offered effective opposition to predatory wealth. The more successful the Forest Service has been in preventing land-grabbing and the absorption of water power by the special interests, the more ingenious, the more devious, and the more dangerous these attacks have become. A favorite one is to assert that the Forest

Service, in its zeal for the public welfare, has played ducks and drakes with the Acts of Congress. The fact is, on the contrary, that the Service has had warrant of law for everything it has done. Not once since it was created has any charge of illegality, despite the most searching investigation and the bitterest attack, ever led to reversal or reproof by either House of Congress or by any Congressional Committee. Not once has the Forest Service been defeated or reversed as to any vital legal principle underlying its work in any court or administrative tribunal of last resort. It is the first duty of a public officer to obey the law. But it is his second duty, and a close second, to do everything the law will let him do for the public good, and not merely what the law directs or compels him to do. Unless the public service is alive enough to serve the people with enthusiasm, there is very little to be said for it.

Another, and unusually plausible, form of attack, is to demand that all land not now bearing trees shall be thrown out of the National Forests. For centuries forest fires have burned through the Western mountains, and much land thus deforested is scattered throughout the National Forests awaiting reforestation. This land is not valuable for agriculture, and will contribute more to the general welfare under forest than in any other way. To exclude it from the National Forests would be no more reasonable than it would be in a city to remove from taxation and municipal control every building lot not now covered by a house. It would be no more reasonable than to condemn and take away from our farmers every acre of land that did not bear a crop last year, or to confiscate a man's winter overcoat because he was not wearing it in July. A generation in the life of a nation is no longer than a season in the life of a man. With a fair chance we can and will reclothe these denuded mountains with forests, and we ask for that chance.

Still another attack, nearly successful two years ago, was an attempt to prevent the Forest Service from telling the people, through the press, what it is accomplishing for them, and how much this Nation needs the forests. If the Forest Service can not tell what it is doing the time will come when there will be nothing to tell. It is just as necessary for the people to know what is being done to help them as to know what is being done to hurt them. Publicity is the essential and indispensable condition of clean and effective public service.

Since the Forest Service called public attention to the rapid absorption of the water-power sites and the threatening growth of a great water-power monopoly, the attacks upon it have increased with marked rapidity. I anticipate that they will continue to do so. Still greater opposition is promised in the near future. There is but one protection — an awakened and determined public opinion. That is why I tell the facts.

CHAPTER XI

THE NEW PATRIOTISM

THE people of the United States are on the verge of one of the great quiet decisions which determine national destinies. Crises happen in peace as well as in war, and a peaceful crisis may be as vital and controlling as any that comes with national uprising and the clash of arms. Such a crisis, at first uneventful and almost unperceived, is upon us now, and we are engaged in making the decision that is thus forced upon us. And, so far as it has gone, our decision is largely wrong. Fortunately it is not yet final.

The question we are deciding with so little consciousness of what it involves is this: What shall we do with our natural resources? Upon the final answer that we shall make to it hangs the success or failure of this Nation in accomplishing its manifest destiny.

Few Americans will deny that it is the manifest destiny of the United States to demonstrate that a democratic republic is the best form of government yet devised, and that the ideals and institutions of the great republic taken together must and do work out in a prosperous, contented, peaceful, and righteous people; and also to exercise, through precept and example, an influence for good among the nations of the world. That destiny seems to us brighter and more certain of realization to-day than ever before. It is true that in population, in wealth, in knowledge, in national efficiency generally, we have reached a place far beyond the farthest hopes of the founders of the Republic. Are the causes which have led to our marvellous development likely to be repeated indefinitely in the future, or is there a reason-

able possibility, or even a probability, that
conditions may arise which will check our
growth?

Danger to a nation comes either from
without or from within. In the first great
crisis of our history, the Revolution, another
people attempted from without to halt the
march of our destiny by refusing to us liberty.
With reasonable prudence and preparedness
we need never fear another such attempt.
If there be danger, it is not from an external
source. In the second great crisis, the
Civil War, a part of our own people strove
for an end which would have checked the
progress of development. Another such
attempt has become forever impossible. If
there be danger, it is not from a division of
our people.

In the third great crisis of our history,
which has now come squarely upon us,
the special interests and the thoughtless
citizens seem to have united together to
deprive the Nation of the great natural

resources without which it cannot endure. This is the pressing danger now, and it is not the least to which our National life has been exposed. A nation deprived of liberty may win it, a nation divided may reunite, but a nation whose natural resources are destroyed must inevitably pay the penalty of poverty, degradation, and decay.

At first blush this may seem like an unpardonable misconception and over-statement, and if it is not true it certainly is unpardonable. Let us consider the facts. Some of them are well known, and the salient ones can be put very briefly.

The five indispensably essential materials in our civilization are wood, water, coal, iron, and agricultural products.

We have timber for less than thirty years at the present rate of cutting. The figures indicate that our demands upon the forest have increased twice as fast as our population.

We have anthracite coal for but fifty

years, and bituminous coal for less than two hundred.

Our supplies of iron ore, mineral oil, and natural gas are being rapidly depleted, and many of the great fields are already exhausted. Mineral resources such as these when once gone are gone forever.

We have allowed erosion, that great enemy of agriculture, to impoverish and, over thousands of square miles, to destroy our farms. The Mississippi alone carries yearly to the sea more than 400,000,000 tons of the richest soil within its drainage basin. If this soil is worth a dollar a ton, it is probable that the total loss of fertility from soil-wash to the farmers and forest-owners of the United States is not far from a billion dollars a year. Our streams, in spite of the millions of dollars spent upon them, are less navigable now than they were fifty years ago, and the soil lost by erosion from the farms and the deforested mountain sides, is the chief reason. The great cattle and sheep ranges of the West, because of overgrazing, are capable, in an average year, of carrying but half the stock they once could support and should still. Their condition affects the price of meat in practically every city of the United States.

These are but a few of the more striking examples. The diversion of great areas of our public lands from the home-maker to the landlord and the speculator; the national neglect of great water powers, which might well relieve, being perennially renewed, the drain upon our non-renewable coal; the fact that but half the coal has been taken from the mines which have already been abandoned as worked out and by caving-in have made the rest forever inaccessible; the disuse of the cheaper transportation of our waterways, which involves comparatively slight demand upon our non-renewable supplies of iron ore, and the use of the rail instead — these are other items in the huge bill of particulars of national waste.

THE FIGHT FOR CONSERVATION

We have a well-marked national tendency to disregard the future, and it has led us to look upon all our natural resources as inexhaustible. Even now that the actual exhaustion of some of them is forcing itself upon us in higher prices and the greater cost of living, we are still asserting, if not always in words, yet in the far stronger language of action, that nevertheless and in spite of it all, they still are inexhaustible.

It is this national attitude of exclusive attention to the present, this absence of foresight from among the springs of national action, which is directly responsible for the present condition of our natural resources. It was precisely the same attitude which brought Palestine, once rich and populous, to its present desert condition, and which destroyed the fertility and habitability of vast areas in northern Africa and elsewhere in so many of the older regions of the world.

The conservation of our natural resources is a question of primary importance on the economic side. It pays better to conserve our natural resources than to destroy them, and this is especially true when the national interest is considered. But the business reason, weighty and worthy though it be, is not the fundamental reason. In such matters, business is a poor master but a good servant. The law of self-preservation is higher than the law of business, and the duty of preserving the Nation is still higher than either.

The American Revolution had its origin in part in economic causes, and it produced economic results of tremendous reach and weight. The Civil War also arose in large part from economic conditions, and it has had the largest economic consequences. But in each case there was a higher and more compelling reason. So with the third great crisis of our history. It has an economic aspect of the largest and most permanent importance, and the motive for action along that line, once it is recognized, should be more than sufficient. But that is not all. In this case, too, there is a higher and more compelling reason. The question of the conservation of natural resources, or national resources, does not stop with being a question of profit. It is a vital question of profit, but what is still more vital, it is a question of national safety and patriotism also.

We have passed the inevitable stage of pioneer pillage of natural resources. The natural wealth we found upon this continent has made us rich. We have used it, as we had a right to do, but we have not stopped there. We have abused, and wasted, and exhausted it also, so that there is the gravest danger that our prosperity to-day will have been bought at the price of the suffering and poverty of our descendants. We may now fairly ask of ourselves a reasonable care for the future and a natural interest in those who are to come after us. No patriotic citizen expects this Nation to run its course and perish in a hundred or two hundred, or five hundred years; but, on the contrary, we expect it to grow in influence and power and, what is of vastly greater importance, in the happiness and prosperity of our people. But we have as little reason to expect that all this will happen of itself as there would have been for the men who established this Nation to expect that a United States would grow of itself without their efforts and sacrifices. It was their duty to found this Nation, and they did it. It is our duty to provide for its continuance in well-being and honor. That duty it seems as though we might neglect — not in wilfulness, not in any lack of patriotic devotion, when once our patriotism is aroused, but in mere thoughtlessness and inability or unwillingness to drop the interests of the moment long enough to realize that what we do now will decide the future of the Nation. For, if we do not take action to conserve the Nation's natural resources, and that soon, our descendants will suffer the penalty of our neglect.

Let me use a homely illustration: We have all known fathers and mothers, devoted to their children, whose attention was fixed and limited by the household routine of daily life. Such parents were actively concerned with the common needs and precautions and remedies entailed in bringing up a family, but blind to every threat that was at all unusual. Fathers and mothers such as these often remain serenely unaware while some dangerous malady or injurious habit is fastening itself upon a favorite child. Once the evil is discovered, there is no sacrifice too great to repair the damage which their unwitting neglect may have allowed to become irreparable. So it is, I think, with the people of the United States. Capable of every devotion in a recognized crisis, we have yet carelessly allowed the habit of improvidence and waste of resources to find lodgment. It is our great good fortune that the harm is not yet altogether beyond repair.

The profoundest duty that lies upon any father is to leave his son with a reasonable equipment for the struggle of life and an untarnished name. So the noblest task that confronts us all to-day is to leave this country unspotted in honor, and unexhausted in resources, to our descendants, who will be, not less than we, the children of the Founders of the Republic. I conceive this task to partake of the highest spirit of patriotism.

CHAPTER XII

THE PRESENT BATTLE

CONSERVATION has captured the Nation. Its progress during the last twelve months is amazing. Official opposition to the conservation movement, whatever damage it has done or still threatens to the public interest, has vastly strengthened the grasp of conservation upon the minds and consciences of our people. Efforts to obscure or belittle the issue have only served to make it larger and clearer in the public estimation. The conservation movement cannot be checked by the baseless charge that it will prevent development, or that every man who tells the plain truth is either a muck-raker or a demagogue. It has taken firm hold on our

national moral sense, and when an issue
does that it has won.

The conservation issue is a moral issue,
and the heart of it is this: For whose benefit
shall our natural resources be conserved —
for the benefit of us all, or for the use and
profit of the few? This truth is so obvious
and the question itself so simple that the
attitude toward conservation of any man in
public or private life indicates his stand in
the fight for public rights.

All monopoly rests on the unregulated
control of natural resources and natural
advantages, and such control by the special
interests is impossible without the help of
politics. The alliance between business and
politics is the most dangerous thing in our
political life. It is the snake that we must
kill. The special interests must get out of
politics, or the American people will put
them out of business. There is no third
course.

Because the special interests are in politics,

we as a Nation have lost confidence in Congress. This is a serious statement to make, but it is true. It does not apply, of course, to the men who really represent their constituents and who are making so fine a fight for the conservation of self-government. As soon as these men have won their battle and consolidated their victory, confidence in Congress will return.

But in the meantime the people of the United States believe that, as a whole, the Senate and the House no longer represent the voters by whom they were elected, but the special interests by whom they are controlled. They believe so because they have so often seen Congress reject what the people desire, and do instead what the interests demand. And of this there could be no better illustration than the tariff.

The tariff, under the policy of protection, was originally a means to raise the rate of wages. It has been made a tool to increase the cost of living. The wool schedule, pro-

[134]

fessing to protect the wool-grower, is found to result in sacrificing grower and consumer alike to one of the most rapacious of trusts.

The cotton cloth schedule was increased in the face of the uncontradicted public testimony of the manufacturers themselves that it ought to remain unchanged.

The Steel interests by a trick secured an indefensible increase in the tariff on structural steel.

The Sugar Trust stole from the Government like a petty thief, yet Congress, by means of a dishonest schedule, continues to protect it in bleeding the public.

At the very time the duties on manufactured rubber were being raised, the leader of the Senate, in company with the Guggenheim Syndicate, was organizing an international rubber trust, whose charter made it also a holding company for the coal and copper deposits of the whole world.

For a dozen years the demand of the

Nation for the Pure Food and Drug bill was outweighed in Congress by the interests which asserted their right to poison the people for a profit.

Congress refused to authorize the preparation of a great plan of waterway development in the general interest, and for ten years has declined to pass the Appalachian and White Mountain National Forest bill, although the people are practically unanimous for both.

The whole Nation is in favor of protecting the coal and other natural resources in Alaska, yet they are still in grave danger of being absorbed by the special interests. And as for the general conservation movement, Congress not only refused to help it on, but tried to forbid any progress without its help. Fortunately for us all, in this attempt it has utterly failed.

This loss of confidence in Congress is a matter for deep concern to every thinking American. It has not come quickly or without good reason. Every man who knows Congress well knows the names of Senators and members who betray the people they were elected to represent, and knows also the names of the masters whom they obey. A representative of the people who wears the collar of the special interests has touched bottom. He can sink no farther.

Who is to blame because representatives of the people are so commonly led to betray their trust? We all are — we who have not taken the trouble to resent and put an end to the knavery we knew was going on. The brand of politics served out to us by the professional politician has long been composed largely of hot meals for the interests and hot air for the people, and we have all known it.

Political platforms are not sincere statements of what the leaders of a party really believe, but rather forms of words which those leaders think they can get others to believe they believe. The realities of the regular political game lie at present far beneath the surface; many of the issues advanced are mere empty sound; while the issues really at stake must be sought deep down in the politics of business — in politics for revenue only. All this the people realize as they never did before, and, what is more, they are ready to act on their knowledge.

Some of the men who are responsible for the union of business and politics may be profoundly dishonest, but more of them are not. They were trained in a wrong school, and they cannot forget their training. Clay hardens by immobility — men's minds by standing pat. Both lose the power to take new impressions. Many of the old-style leaders regard the political truths which alone insure the progress of the Nation, and will hereafter completely dominate it, as the mere meaningless babble of political infants. They have grown old in the belief that money has the right to rule, and they can never

[138]

understand the point of view of the men who recognize in the corrupt political activity of a railroad or a trust a most dangerous kind of treason to government by the people.

When party leaders go wrong, it requires a high sense of public duty, true courage, and a strong belief in the people for a man in politics to take his future in his hands and stand against them.

The black shadow of party regularity as the supreme test in public affairs has passed away from the public mind. It is a great deliverance. The man in the street no longer asks about a measure or a policy merely whether it is good Republican or good Democratic doctrine. Now he asks whether it is honest, and means what it says, whether it will promote the public interest, weaken special privilege, and help to give every man a fair chance. If it will, it is good, no matter who proposed it. If it will not, it is bad, no matter who defends it.

[139]

It is a greater thing to be a good citizen than to be a good Republican or a good Democrat.

The protest against politics for revenue only is as strong in one party as in the other, for the servants of the interests are plentiful in both. In that respect there is little to chose between them.

Differences of purpose and belief between political parties to-day are vastly less than the differences within the parties. The great gulf of division which strikes across our whole people pays little heed to fading party lines, or to any distinction in name only. The vital separation is between the partisans of government by money for profit and the believers in government by men for human welfare.

When political parties come to be badly led, when their leaders lose touch with the people, when their object ceases to be everybody's welfare and becomes somebody's profit, it is time to change the leaders. One of the most significant facts of the time is that the professional politicians appear to be wholly unaware of the great moral change which has come over political thinking in the last decade. They fail to see that the political dogmas, the political slogans, and the political methods of the past generation have lost their power, and that our people have come at last to judge of politics by the eternal rules of right and wrong.

A new life is stirring among the dry bones of formal platforms and artificial issues. Morality has broken into politics. Political leaders, Trust-bred and Trust-fed, find it harder and harder to conceal their actual character. The brass-bound collar of privilege has become plain upon their necks for all men to see. They are known for what they are, and their time is short. But when they come to be retired it will be of little use to replace an unfaithful public servant who wears the collar by another public servant with the same collar around his neck. Above all, what we need in every office is free men representing a free people.

The motto in every primary — in every election — should be this: No watch-dogs of the Interests need apply.

The old order, standing pat in dull failure to sense the great forward sweep of a nation determined on honesty and publicity in public affairs, is already wearing thin under the ceaseless hammering of the progressive onset. The demand of the people for political progress will not be denied. Does any man, not blinded by personal interest or by the dust of political dry rot, suppose that the bulk of our people are anything else but progressive? If such there be, let him ask the young men, in whose minds the policies of to-morrow first see the light.

The people of the United States demand a new deal and a square deal. They have grasped the fact that the special interests are now in control of public affairs. They

have decided once more to take control of
their own business. For the last ten years
the determination to do so has been swelling
like a river. They insist that the special
interests shall go out of politics or out of
business — one or the other. And the choice
will lie with the interests themselves. If
they resist, both the interests and the people
will suffer. If wisely they accept the inevitable, the adjustment will not be hard. It
will do their business no manner of harm to
make it conform to the general welfare.
But one way or the other, conform it must.

The overshadowing question before the
American people to-day is this: Shall the
Nation govern itself or shall the interests
run this country? The one great political
demand, underlying all others, giving meaning to all others, is this: The special interests must get out of politics. The old-style
leaders, seeking to switch public attention
away from this one absorbing and overwhelming issue are pitifully ridiculous and

out of date. To try to divert the march of
an aroused public conscience from this
righteous inevitable conflict by means of
obsolete political catchwords is like trying
to dam the Mississippi with dead leaves.

To drive the special interests out of politics
is a vast undertaking, for in politics lies their
strength. If they resist, as doubtless they
will, it will call for nerve, endurance, and
sacrifice on the part of the people. It will
be no child's play, for the power of privilege
is great. But the power of our people is
greater still, and their steadfastness is equal
to the need. The task is a tremendous one,
both in the demands it will make and the
rewards it will bring. It must be under-
taken soberly, carried out firmly and justly,
and relentlessly followed to the very end.
Two things alone can bring success. The
first is honesty in public men, without which
no popular government can long succeed.
The second is complete publicity of all the
affairs in which the public has an interest,

such as the business of corporations and political expenses during campaigns and between them. To these ends, many unfaithful public servants must be retired, much wise legislation must be framed and passed, and the struggle will be bitter and long. But it will be well worth all it will cost, for self-government is at stake.

There can be no legislative cure-all for great political evils, but legislation can make easier the effective expression and execution of the popular will. One step in this direction, which I personally believe should be taken without delay, is a law forbidding any Senator or Member of Congress or other public servant to perform any services for any corporation engaged in interstate commerce, or to accept any valuable consideration, directly or indirectly, from any such corporation, while he is a representative of the people, and for a reasonable time thereafter. If such a law would be good for the Nation in its affairs, a similar law should be good for the States and the cities in their affairs. And I see no reason why Members and Senators and State Legislators should not keep the people informed of their pecuniary interest in interstate or public service corporations, if they have any. It is certain such publicity would do the public no harm.

This Nation has decided to do away with government by money for profit and return to the government our forefathers died for and gave to us — government by men for human welfare and human progress.

Opposition to progress has produced its natural results. There is profound dissatisfaction and unrest, and profound cause for both. Yet the result is good, for at last the country is awake. For a generation at least there has not been a situation so promising for the ultimate public welfare as that of to-day. Our people are like a hive of bees, full of agitation before taking flight to a better place. Also they are ready to sting. Out of the whole situation shines the confident hope of better things. If any man is discouraged, let him consider the rise of cleaner standards in this country within the last ten years.

The task of translating these new standards into action lies before us. From sea to sea the people are taking a fresh grip on their own affairs. The conservation of political liberty will take its proper place alongside the conservation of the means of living, and in both we shall look to the permanent welfare by the plain people as the supreme end. The way out lies in direct interest by the people in their own affairs and direct action in the few great things that really count.

What is the conclusion of the whole matter? The special interests must be put out of politics. I believe the young men will do it.

INDEX

14 DAY USE
RETURN TO DESK FROM WHICH BORROWED
LOAN DEPT.

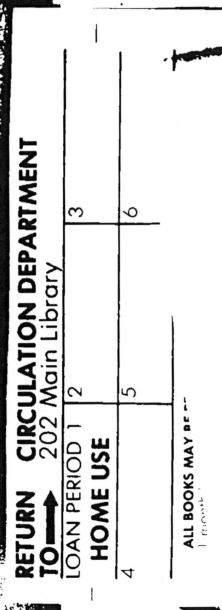

RETURN TO ➔ CIRCULATION DEPARTMENT
202 Main Library

LOAN PERIOD 1 HOME USE	2	3
4	5	6

ALL BOOKS MAY BE

LaVergne, TN USA
17 August 2010
193649LV00005B/64/P